EARTH'S FOUR SPHERES

The Biosphere

By Karen McMichael

Cavendish Square

New York

Published in 2023 by Cavendish Square Publishing, LLC
29 E. 21st Street, New York, NY 10010

Website: cavendishsq.com

Library of Congress Cataloging-in-Publication Data

Names: McMichael, Karen, author.
Title: The biosphere / Karen McMichael.
Description: New York : Cavendish Square, [2023] | Series: The inside guide: earth's four spheres | Includes index.
Identifiers: LCCN 2021060954 | ISBN 9781502664457 (set) | ISBN 9781502664464 (library binding) | ISBN 9781502664440 (paperback) | ISBN 9781502664471 (ebook)
Subjects: LCSH: Biosphere–Juvenile literature.
Classification: LCC QH343.4 .M38 2023 | DDC 577–dc23/eng/20211227
LC record available at https://lccn.loc.gov/2021060954

Editor: Kate Mikoley
Copyeditor: Abby Young
Designer: Deanna Paternostro

The photographs in this book are used by permission and through the courtesy of: Cover Dmitry Rukhlenko/Shutterstock.com; p. 4 HelloRF Zcool/Shutterstock.com; p. 6 Bill45/Shutterstock.com; p. 7 pst.rtw/Shutterstock.com; p. 8 herle_catharina/Shutterstock.com; p. 9 danylyukk1/Shutterstock.com; p. 10 Romolo Tavani/Shutterstock.com; p. 12 AustralianCamera/Shutterstock.com; p. 13 VectorMine/ Shutterstock.com; p. 14 alinabel/Shutterstock.com; p. 15 Ryan M. Bolton/Shutterstock.com; p. 16 Dashu Xinganling/Shutterstock.com; p. 17 sittitap/Shutterstock.com; p. 18 (top) TSN52/Shutterstock.com; p. 18 (bottom) geogif/Shutterstock.com; p. 20 (top) Aostojska/ Shutterstock.com; p. 20 (bottom) Marek Rybar/Shutterstock.com; p. 21 Irina Markova/Shutterstock.com; p. 22 Artsiom P/ Shutterstock.com; p. 24 herain kanthatham/Shutterstock.com; p. 25 Mike Workman/Shutterstock.com; p. 26 shao weiwei/Shutterstock.com; p. 27 narikan/Shutterstock.com; p. 28 (top) Nikitin Victor/Shutterstock.com; p. 28 (bottom) RUKSUTAKARN studio/Shutterstock.com; p. 29 (top) Traveller70/Shutterstock.com; p. 29 (bottom) Maples Images/Shutterstock.com.

CPSIA compliance information: Batch #CSCSQ23: For further information, contact Cavendish Square Publishing LLC, New York, New York, at 1-877-980-4450.

Printed in the United States of America

CONTENTS

Living things, or organisms, make Earth a unique planet! Organisms include plants, animals, bacteria, and fungi.

A LIVELY PLANET

Pictures of Earth from space show its swirling cloud patterns, bright white poles, blue-black oceans, and sandy deserts. These different features of Earth are all beautiful in their own way. However, it's not just its beauty that makes Earth special—it's also the only planet in our solar system known to support life!

Earth's Spheres

Everything on Earth can be placed into one of four spheres, or the systems in which Earth's many processes occur. While each sphere is unique, they all work together to keep Earth functioning normally.

The atmosphere is the whole mass of air that surrounds Earth. The lithosphere is the solid part of Earth, or its land. The hydrosphere is the water on Earth in all its forms.

Bio comes from the Greek word for "life." "Biosphere" is the sphere of life, and "biology" is the study of life.

This book is about the biosphere—the part of Earth in which life can exist. Within the biosphere, living creatures—

The lithosphere, hydrosphere, atmosphere, and biosphere constantly interact to make life on Earth possible.

from tiny, single-celled organisms to huge mammals—constantly interact with their **environments** in a never-ending cycle of life and death. You're alive—that means you're part of the biosphere!

The Early Earth

Our planet first formed about 4.5 billion years ago. Scientists estimate that life appeared on Earth more than 3.7 billion years ago. These life-forms were microbes, or extremely small living things that can only be seen with a microscope. From these life-forms came other life-forms, such as bacteria, dinosaurs, and human beings.

In Greenland, there are fossils of cyanobacteria, or blue-green algae, that are older than 3.7 billion years old. These are called stromatolites and are some of the oldest signs of life existing on Earth.

Even though scientists have a large amount of information about when life on Earth may have first appeared, it's hard to be exactly sure

Fast Fact

The Greenland fossils might not be from the first life-forms on Earth. Some scientists think that because these organisms are so complex, life must have actually started well before 3.7 billion years ago.

Stromatolites, such as those shown here, still form today.

what processes took place to make life possible. Much about early life on Earth is still a mystery today.

Spheres Working Together

Earth's four spheres are interdependent, meaning they work together and changes in one sphere can cause changes in the others. They interact through processes that are biotic, or living, and abiotic, or nonliving. These processes keep food and energy in motion. Just as the water cycle takes place mainly in the hydrosphere and atmosphere, the biosphere features a number of unique processes. The oxygen cycle, the nitrogen cycle, and other cycles constantly move life-giving elements through the biosphere.

Fast Fact

Weather is the state of the air and atmosphere at a particular time and place. Climate is the average weather conditions of a place over a period of years.

The changing of seasons, weather, and climate are examples of abiotic processes that affect the biosphere. These are just a few of the factors that influence the biosphere's **biota**, creating a **dynamic** biosphere, supporting a wide variety of living things.

The changing of seasons is an abiotic process, but it has a big impact on the biosphere's biotic processes.

THE CARBON CYCLE

Carbon is often described as "the building block of life." Every living thing—and many nonliving things—are made up of carbon. This life-giving element is constantly moving within the biosphere through the carbon cycle. Most plants grow through a process called photosynthesis, which involves absorbing, or taking in, carbon dioxide (CO_2) from Earth's atmosphere. Animals take in carbon by eating plants or other animals. When plants and animals die, carbon dioxide is returned to the biosphere to keep the cycle in motion. Carbon dioxide is also released into the atmosphere when animals breathe and when humans burn fossil fuels, which are made of things that were once living.

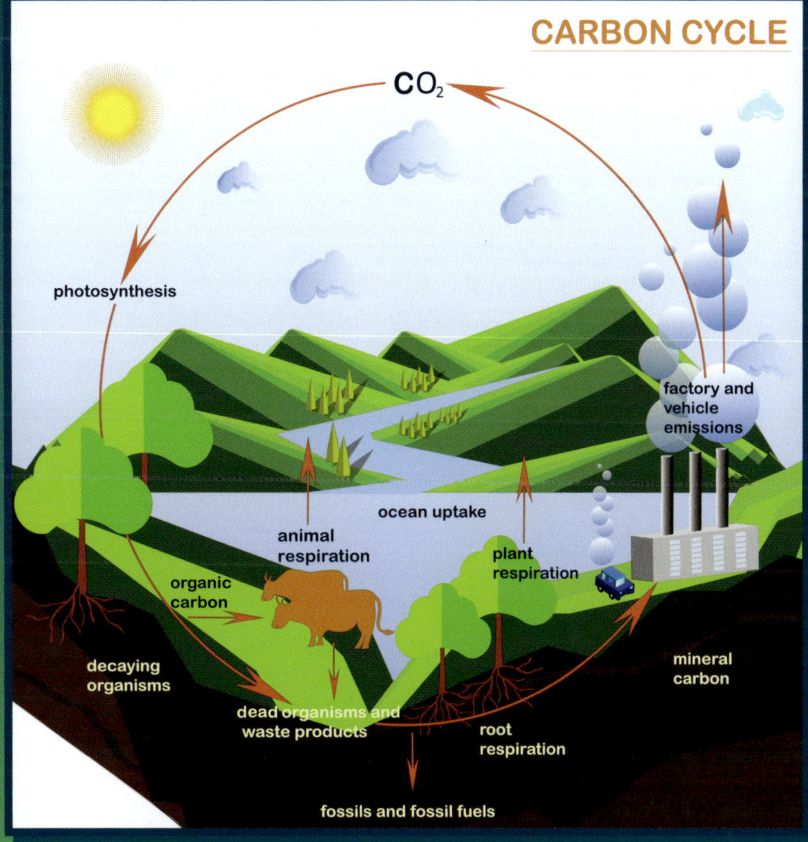

All living things are built on carbon. The carbon cycle helps keep Earth in balance and stops the planet's carbon from all going into the atmosphere.

Without the sun, none of the plants or animals that make our planet unique could survive!

THE BASICS OF LIFE ON EARTH

Organisms need energy to survive. Energy for all life on Earth can be traced back to the sun. In fact, some might even say the sun is what gives everything on Earth life. Learning the basics of life on Earth begins with the sun.

The Amazing Sun

The sun gives all living things energy through photosynthesis, the process by which green plants and a few other organisms turn water and carbon dioxide into a kind of food in the presence of sunlight.

Everything in the biosphere fits into at least one food chain, which is comprised of producers, consumers, and decomposers. Producers, also known as autotrophs, make their own food through photosynthesis. The only organisms that do this are plants, algae, phytoplankton, and some bacteria. Producers are the first organisms in all food chains.

Fast Fact

Animals that eat both plants and animals, such as many secondary consumers, are known as omnivores.

Primary consumers called herbivores get their energy from consuming producers. Secondary consumers

Producers, such as plants, use the sun to make their own food. Then, they serve as food for other organisms in their food chain.

get their energy from consuming primary consumers and producers. Tertiary consumers get their energy from eating secondary consumers. Decomposers get their energy from decomposing, or breaking down, dead plant and animal matter.

It might seem like only plants and other organisms that make food through photosynthesis rely on this process. However, without photosynthesis, the other organisms in the biosphere wouldn't have any food! Producers need the sun, and other organisms need producers.

Organisms don't just rely on the food they eat to survive. They need all the organisms that come before them in the food chain too. Otherwise, the food they eat wouldn't be available!

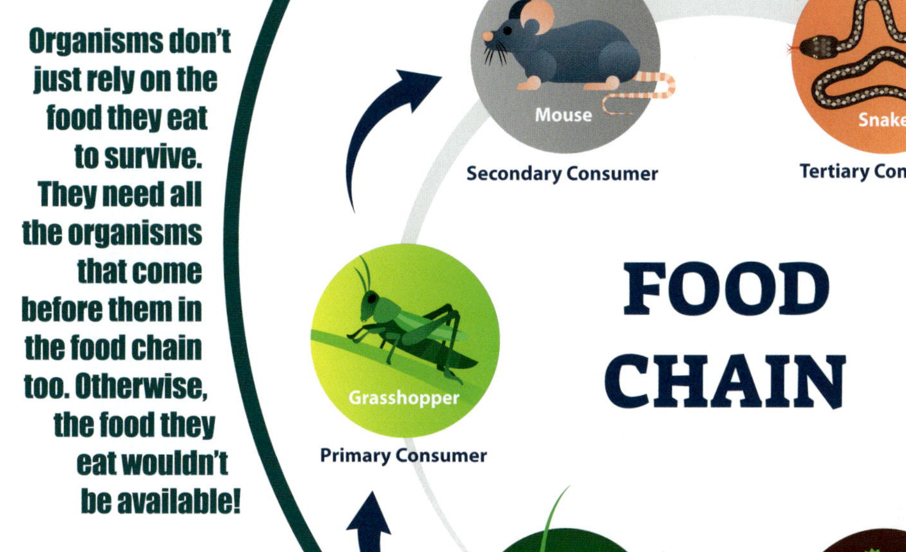

Evolution and Natural Selection

An English naturalist named Charles Darwin shook the world when he published his theories on **evolution** in 1859. In his book *On the Origin of Species*, Darwin explained how species can change over time through a process called natural selection. This process, also known as "survival of the fittest," means that all species must adapt to fit their environment, and if they don't, they'll go extinct, or die out.

Fast Fact

When a number of food chains interact, they form a food web.

Organisms that are better adapted to their environment are the ones that reproduce successfully. Then, they pass on the desired **traits** to their

TROPHIC LEVELS AND ENERGY PYRAMIDS

The organisms in a food chain can be further separated into trophic levels. These are steps based on an organism's feeding behavior. Autotrophs are the first level, herbivores are the second level, and carnivores and omnivores are the third level. Sometimes the levels are separated into smaller, more detailed levels.

Energy pyramids describe how energy is lost from one trophic level to the next. As organisms are consumed, only about 10 percent of the energy from that organism is transferred, or given over, to the organism at the next level.

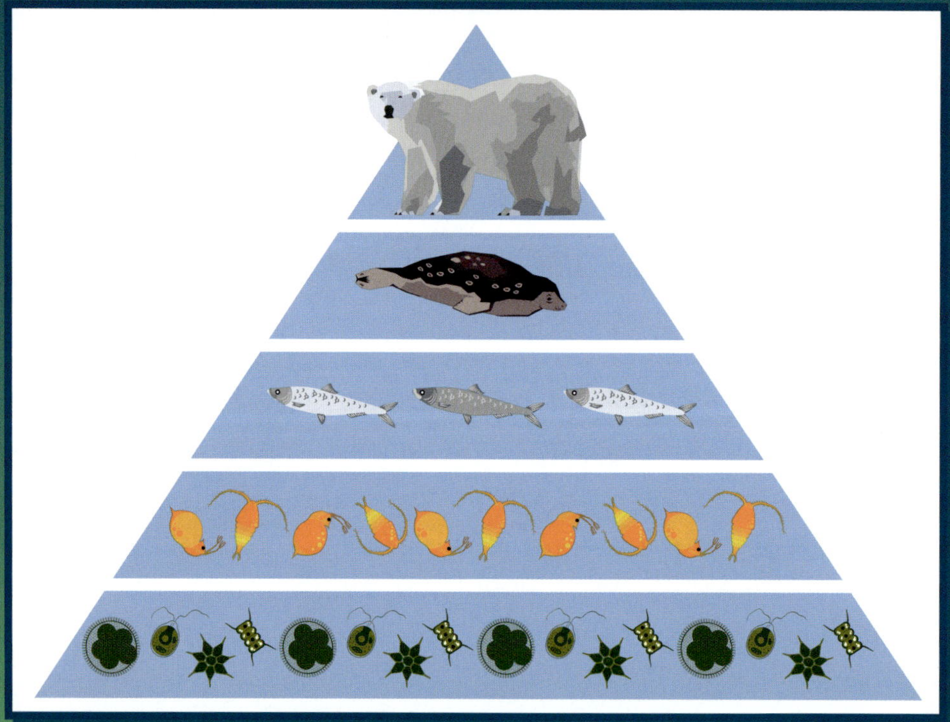

A pyramid gets smaller as it nears the top. Energy is lost when an organism is eaten. The smaller parts of the pyramid reflect smaller amounts of energy.

Fast Fact

Darwin developed his theories, or ideas, by studying life on islands, like the Galápagos Islands off the coast of Ecuador. To Darwin, islands were living laboratories where evolution could be seen in real time.

Darwin studied the beaks of finches on the Galápagos Islands. Scientists still study these birds today! After a drought in 2003, scientists noted their beaks became stronger, allowing them to eat tougher seeds.

offspring. The organisms that are less fit for their environment do not reproduce as quickly as other species and eventually die out.

One example of this can be seen with mice. White mice are easier for predators to see on the dark forest floor. Therefore, the white mice are eaten by hawks first. If there are no white mice left to reproduce, the trait for white fur won't be passed on to the next generation.

Evolution by natural selection has created some amazing animal and plant adaptations. Natural selection can also cause species to behave in certain ways. For example, many animals have developed certain mating rituals, or behaviors, due to natural selection.

Biomes change over time. It's hard to believe, but 10,000 years from now, a lush grassland such as this one could be a desert!

ALL ABOUT BIOMES

Earth's surface can be separated into different communities, called biomes, based on climate, types of soil, and the species of plants and animals that live there. The five major biomes are forest, desert, tundra, grassland, and aquatic. Each biome can be further separated into smaller **ecosystems**.

> **Fast Fact**
>
> The Sahara wasn't always as large a desert as it is today. Parts of it used to be lush and green. However, the climate changed, and these areas dried out and became covered with sand.

Types of Forests

Found near the equator, tropical forests are known for having only two seasons—rainy and dry. The temperature remains fairly constant—usually 68 to 77 degrees Fahrenheit (20 to 25 degrees Celsius) all year—and more than 78 inches (2,000 millimeters) of rain

In tropical forests, the growing seasons are long with periods of heavy rainfall.

Unlike in tropical forests, the canopies in temperate forests allow light to reach the forest floor.

falls annually. Lush canopies prevent light from reaching the forest floor. Many plants and animals live in tropical forests. Temperate forests, which exist in eastern North America, northeastern Asia, and western and central Europe, experience all seasons. The temperature varies greatly, between -22 and 86° F (-30 and 30° C). These forests receive far less precipitation than tropical forests.

Fast Fact

Soil in tropical forests is loose and often lacks nutrients due to heavy rains. The process of soil losing **nutrients** due to heavy rain is called leaching.

Deserts

Deserts are large, dry biomes with little vegetation. They are often classified as receiving less than 10 inches (25 centimeters) of rainfall per year. The

The desert can seem like a harsh biome for organisms, but those that live there have adapted to do well in such an environment.

THE TAIGA

The taiga, also known as the boreal forest region, is the largest land biome. It is a kind of forest that exists just south of the arctic tundra and is characterized by short, wet, and somewhat warm summers and long, cold, and dry winters. The growing season in the taiga is only 130 days.

The taiga is known for its many conifer, or cone-bearing, trees. Loggers often remove trees from taiga regions. This can be harmful to the biome and can cause certain species of plants and animals to go extinct due to habitat loss.

plants and animals in deserts have adapted to living there. Different types of deserts include hot and dry (such as the Sahara), semiarid (such as the Australian Outback), coastal (such as the Atacama Desert), and cold (such as the entire continent of Antarctica).

Grasslands

Grassland biomes are large areas covered mostly by grasses rather than trees. They're divided into tropical grasslands, called savannas, and temperate grasslands. Grasslands are home to large, migrating herd animals, such as elephants in Africa and pronghorns in North America.

Grasslands are often used for farming, but their soil can be stripped of nutrients if the same crops are grown on the land year after year. When soil lacks nutrients, crops can't be grown there for a while. Also, in areas with lots of farming, many grassland animals have been wiped out.

The African savanna, shown here, is an example of a tropical grassland.

The Tundra

Tundra biomes have few trees, very low temperatures, little precipitation, and **permafrost**. The two types of tundra are arctic and alpine.

The arctic tundra circles the North Pole and reaches down to the taiga. Its growing season is very short at just

Tundra biomes are known for having cycles in population sizes. For example, if the population of lemmings grows too large, groups of lemmings will migrate to find less crowded places to live.

50 to 60 days. The plants there have adapted to strong winds and cold temperatures.

The alpine tundra exists on mountains at elevations above which trees can't grow. The growing season is about 180 days.

Fast Fact

The aquatic biome is the largest biome, covering about 75 percent of Earth's surface.

Life in the Aquatic

The aquatic biome can be broken down into freshwater regions and marine regions. Ponds, lakes, rivers, streams, and freshwater wetlands are freshwater regions. Oceans, coral reefs, and estuaries are marine regions. The oceans are the largest of these regions. Most marine life occurs near the surface of the ocean where sunlight can be seen, which is called the photic zone.

Coral reefs are often called the "rainforests of the oceans" because they are home to most of Earth's marine species.

All of Earth's systems work together to sustain the planet we call home. To keep the biosphere healthy, we need to keep all of Earth's spheres healthy.

ONE BIOSPHERE, ONE PLANET

Earth has been around for more than 4.5 billion years. Some life-forms have been around for close to 3.7 billion years. However, human beings have been around for only 1.8 million years or so. These early humans, called *Homo erectus*, are the earliest creatures showing similarities to present-day humans, or *Homo sapiens*. It may seem like humans have been around a long time, but in the history of the world, it's relatively short. Still, we've had a large impact on the planet. It's up to us to protect it.

Humans of Today

Homo sapiens evolved in Africa around 300,000 years ago. These humans, like other species before them, were hunters and gatherers. They evolved to adapt to their surroundings. They used specialized tools to help them deal with their environments. The brains of *Homo sapiens* are much larger than our ancestors' brains, allowing us to learn from and interact with each other in complex ways.

Fast Fact

Humans are primates, and the closest relatives we have are chimpanzees and bonobos. Together, we all fit into the great ape family, Hominidae.

ECOSYSTEM SERVICES

Ecosystem services are the benefits humans get from a healthy biosphere. They can be broken down into four categories: provisioning, regulating, habitat, and cultural services. Provisioning services are products from ecosystems to keep life going, such as food and medicine. Regulating services are when Earth works to keep ecosystems going through processes such as water purification and pollination. Habitat services are when ecosystems provide habitats to ensure the passage of traits through offspring. Cultural services include ways humans enjoy ecosystems, such as by participating in outdoor sports or admiring nature-themed art.

Bees may be small creatures, but they play a huge role in the biosphere. They pollinate most flowering plants. Without them, much of the fruits, nuts, and vegetables grown in the United States would likely die out.

Fast Fact

Many of the medicines we use come from organisms in our biosphere.

As a result, humans have been able to move to the far reaches of Earth and grow in number. There are now 7.8 billion of us.

Protecting Biodiversity

Every organism in the biosphere plays an important role, or part, in its ecosystem. This **biodiversity** creates a stable and healthy biosphere. Preserving Earth's biodiversity is more important now than ever.

Many scientists believe we are experiencing another great extinction, the likes of which Earth hasn't seen since the dinosaurs vanished about 65 million years ago. Today, Earth is losing species—and losing biodiversity—at an alarming rate. All the food on Earth comes directly from the plants and animals within the biosphere. Extinction is permanent. We must be careful that we aren't treating the biosphere wastefully.

When we lose biodiversity, we risk upsetting the delicate balance that keeps the ecosystems within our biosphere functioning properly.

Sustaining the Biosphere

Earth's climate is changing, and most scientists agree it's being driven mainly by humans' overuse of **fossil fuels**. As a result, the polar ice caps are melting and oceans are warming, leading to more frequent and powerful storms. Earth is becoming drier in many locations, deserts

Large-scale desalination plants can change undrinkable salt water into fresh water for human use. However, these plants are expensive to build and require lots of energy to run.

are expanding, and stores of fresh water are dropping to critical levels. Plants and animals will have to adapt or risk going extinct.

Humans also need to adapt. We need to find new ways to protect and manage the natural resources—the nutrient-rich soil, fresh water, plants, and animals—we need for our biosphere to function properly. In doing so, we can help sustain the biosphere for future generations.

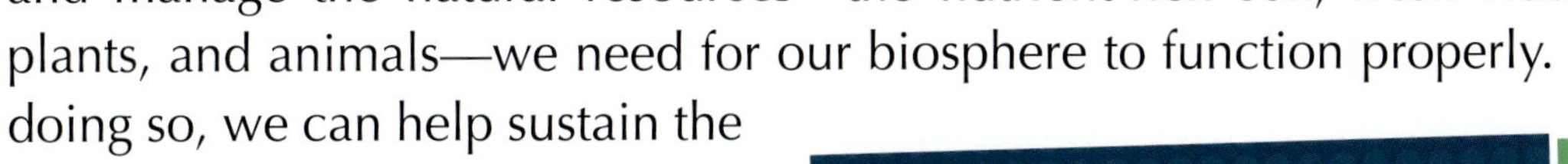

Fast Fact

Habitat loss is one of the greatest threats to Earth's biodiversity. When we cut down forests or build cities, we destroy the habitats in which plants and animals live.

Earth's population is the largest it's ever been. The biosphere has to support all humans and all other living things. To ensure Earth continues to support us, it's necessary to do things to take care of it, such as recycling and planting trees.

THINK ABOUT IT!

1. The biosphere can only exist with the help of the other spheres. What are some ways the hydrosphere, lithosphere, and atmosphere help the world's organisms stay alive?

2. We know consumers need producers to survive, but can you think of any ways producers rely on consumers?

3. Imagine we didn't have modern technology. Which biome do you think would be best for supporting human life? Why?

4. What's one thing you could do to help the biosphere today?

GLOSSARY

biodiversity: The existence of many different kinds of plants and animals in an environment.

biota: The plants and animals of a region.

dynamic: Always active or changing.

ecosystem: All the living things in an area.

environment: The conditions that surround a living thing and affect the way it lives.

evolution: The process of change in an animal or a plant species over long periods of time.

fossil fuel: A fuel—such as coal, oil, or natural gas—that is formed in the earth from dead plants or animals.

nutrient: Something taken in by a plant or animal that helps it grow and stay healthy.

permafrost: A layer of soil that is always frozen.

trait: A quality that makes one person or thing different from another.

unique: Special or different from anything else.

FIND OUT MORE

Books

Claybourne, Anna. *Make a Biosphere and Mini Garden*. New York, NY: Crabtree Publishing Company, 2020.

Gregory, Joy. *Biosphere*. New York, NY: Lightbox, 2022.

Perdew, Laura. *Biodiversity: Explore the Diversity of Life on Earth with Environmental Science Activities for Kids*. White River Junction, VT: Nomad Press, 2019.

Websites

About the Biosphere
mynasadata.larc.nasa.gov/basic-page/about-biosphere
Watch a short video and learn more about the biosphere on this website from NASA.

Biosphere
www.nationalgeographic.org/encyclopedia/biosphere/
Read more about Earth's biosphere here.

Biosphere Facts
www.softschools.com/facts/earth_systems/biosphere_facts/3229/
Go to this website to find out more about the biosphere.

Publisher's note to educators and parents: Our editors have carefully reviewed these websites to ensure that they are suitable for students. Many websites change frequently, however, and we cannot guarantee that a site's future contents will continue to meet our high standards of quality and educational value. Be advised that students should be closely supervised whenever they access the internet.

INDEX